The Systems Thinking Child

A book series, sponsored by the Institute for Integrated Systems Thinking, invites young readers into a world where everything is beautifully connected. Rooted in a teaching philosophy that nurtures empathy, curiosity, and deep respect for all life, this series encourages children to explore how humans, animals, nature, and energy systems are all part of one vibrant, interwoven whole. Through engaging storytelling and rich illustrations, each book helps young minds build a sense of belonging in the world—and a responsibility to care for it.

In this heartwarming and educational story, a wise grandmother shares with her curious granddaughter the incredible role the Sun plays in sustaining life on Earth. Through gentle storytelling and vivid imagery, children will learn how the Sun helps plants grow, keeps animals warm, and lights up our world every day. This beautifully illustrated tale blends science and love, inspiring young readers to appreciate the wonders of nature and the warmth of family.

This is a multilingual edition.

Fortuna is a curious little girl who lives in a small town called Midland. She lives in a cozy house with her mom, dad, big sister, baby brother, and her grandma, Hikma.

Before she retired, Grandma Hikma was an astronomy teacher. She loved teaching, and she still does! Now, she teaches her grandkids, including Fortuna.

Fortuna loves to play outside and look around. She likes watching birds fly, bees buzz, and flowers bloom.

One day, Fortuna asked, "Granny, where do all the pretty flowers come from?" Grandma Hikma smiled and said, "That's a great question! Most flowers come from tiny seeds that take a nap in the ground.
But guess what? They need something magical to wake up and grow!"

"Let me guess," said Fortuna. "A giant bell wakes them up!" Granny giggled. "Not quite! Try again!"

"I know!" said Fortuna with a big smile. "A squirrel digs up the seeds in winter and gives them a hug!"

"What would hungry squirrels do when
they find seeds?" asked Granny.
"They would eat them!" said Fortuna,
rolling her eyes a little.
Then she asked, "So, tell me, Granny?"

"Shams!" Grandma said.
"Who's that?" asked
Fortuna.
Grandma smiled and said,
"I'll give you some hints,
then you guess—because I
know you already know
Shams!"
She leaned back in her chair,
getting comfy.

"Shams is bright and hot,
loves to wake up and go to
bed,
and sometimes plays hide-
and-seek behind the
clouds," Granny said with a
wink.

9

"Is it our neighbor, Mr. Slickshine?" said Fortuna. "He always says he's too hot, loves wearing shiny clothes, can't decide if he should sit or stand when he talks, and sometimes hides behind bushes to surprise us!"

"Very funny, Fortuna, but it's not Mr. Slickshine," said Granny.
"Want to try again?" Granny asked.
"No, Granny, I want you to tell me!" said Fortuna.

"There," said Granny, pointing to the sun high up in the sky.

"The sun is like a big, warm hug for the seeds.
It helps them grow tall and strong, just like you do when you eat yummy food."

Fortuna looked up at the bright, shining sun. "Shams?" she asked. "The sun is Shams, and it helps flowers grow?"

"Absolutely!" Granny said
with a smile.
"The flowers make sweet
nectar for the bees.
Bees love nectar! They fly
from flower to flower,
collecting it to make
yummy honey.
And guess what, Fortuna?
We love honey on our toast
and in our tea!"

Fortuna's eyes sparkled.
"So, flowers help bees, and bees help us?"
"Exactly!" Granny said with a big smile.
"It's like a magical game of passing the ball!"

"The sun helps the flowers, the flowers help the bees, and the bees help us," said Granny.

She sank deeper into her rocking chair and started to tell stories about Shams.

Fortuna thought, "Granny and Shams must be old friends, since they're both so old!"

17

Granny said,
"Shams is the big, bright ball of energy. Shams lives far, far away, but Shams is with us all the time. We all need Shams.
Shams' special rays, called sunshine, are like a magic bond that connects all of us. Shams helps plants grow, and plants make yummy food for us and all animals."

Imagine a world without Shams, and you would imagine a world without yummy fruits and vegetables!
But Shams does not just hug the seeds in the ground to wake them up and help them grow.
Shams does many other important things for the seeds and for everything else!

There's more magic!
Shams helps make
electricity too!
We use special panels to
catch Shams's rays and
turn them into electricity.
Isn't that amazing?

So, Fortuna, the next time you turn on a light or watch your favorite show, remember the warm, sunny rays. They helped bring you that magic energy!

Now, when you eat yummy
fruits or vegetables,
you're like eating some of
Shams's energy!
It's like the plants stored the
sun's power inside them,
and when you eat them, you get
that power to do cool things like
run and jump.

Even the bread in your
sandwich,
the cheese on your pizza,
and the milk in your cereal
all come from plants that got
their energy from Shams!

So, every time you eat
something,
you're getting a little bit of
Shams's Light.

"I'm grateful to Shams !"
she said to her Granny, who
was now dozing off, napping
under the warm sunshine.

"We all are, Fortuna,"
Grandma replied, barely
opening her eyes.
"You're a very special part of
our world, and the world is
in you."

Fortuna smiled. She knew that even though she was small, she was part of something truly and amazingly connected.

She turned to her
Granny to give her a
hug. Granny was
deep asleep, snoring
like an out-of-tune
bass.
Fortuna hugged her
anyway and sat
beside her, soaking
in the warm rays of
sunshine.

end